# 101 Ways To Use AI In Everyday Life

By

Larry Richman

Century Publishing

Salt Lake City, Utah

*101 Ways to Use AI in Everyday Life* by Larry Richman



ISBN: 978-0-941846-42-4



centurypubl.com

info@centurypubl.com

The author used AI to help generate and organize some of the ideas for this book, but the writing and editing is his.

Printed in the United States of America.

# Contents

# Introduction

We live in a time when artificial intelligence (AI) is a part of daily life. AI helps computers and machines simulate human intelligence and problem-solving. It is changing the way we work, learn, and connect. It has never been easier to create, build a business, or improve how we do everyday tasks.

Artificial intelligence (AI) is a tool designed to amplify human creativity and ingenuity, not to replace it. It is a remarkable tool that "offers the potential of advancing knowledge, improving our quality of life, facilitating communication and connection, enhancing personal learning and growth, and fostering creativity and innovation." (Bednar, David A., Young Adult Devotional, November 3, 2024)

## History of Technology

Technology has always shaped the way people live and work. The Industrial Revolution in the 1700s and 1800s brought machines, steam power, and factories. These inventions made work faster and more efficient. The invention of the automobile in the late 1800s changed how people traveled. The telephone changed how people talked. Then, in the 1990s, the internet connected billions of people and things around the world and made information easy to find. These innovations collectively reshaped education, business, and global connectivity.

Now, AI is the next big step. It powers voice assistants, medical diagnostics, and even self-driving cars. AI can find patterns, make predictions, and solve problems in fast and efficient ways.

## Principles For the Good and Uplifting Use of AI

AI can bless your life when used with wisdom and purpose, but it can also create risks if used carelessly. When used thoughtfully, AI can be a powerful companion in many aspects of your life.

Consider the following principles that can provide guardrails in your use of AI.

1. ***Don't consider AI as an infallible source of truth.*** Use AI to gather, organize, and summarize information, but always verify information with multiple trustworthy sources whose information proves to be consistently reliable. AI tools are trained on large language models (LLMs) built from vast collections of online text that inevitably include errors, misinformation, and even outright falsehoods. Also, although AI tools are improving rapidly, they sometimes "hallucinate," or give you information that sounds plausible but is misleading or entirely made up.

2. ***Don't let AI judge for you.*** Gather information with AI, but don't let it replace your own thinking, judgment, or moral reasoning. "Artificial intelligence is not a substitute for human intelligence; it is a tool to amplify human creativity and ingenuity" (Fei-Fei Li, Co-Director of the Stanford Institute for Human-Centered Artificial Intelligence). Outsourcing interpretation to AI tools weakens your ability to discern truth and act according to your personal and spiritual values.

3. ***Use AI to uplift and contribute to the common good.*** Employ AI tools to create text, images, videos, music, or ideas that encourage virtue, strengthen community, and elevate the moral tone of society. Let your use of AI reflect a commitment to goodness, integrity, and service.

4. ***Don't humanize AI.*** Remember that AI is software and computation—not a person. AI can mimic empathy, but it cannot love, care, or understand you. Avoid treating AI as a companion, confidant, or emotional substitute that displaces real human relationships.

5. ***Watch out for deepfakes.*** AI can generate convincing but false audio, images, and video. Always verify with multiple trustworthy sources before accepting or sharing information. Realism does not guarantee authenticity.

6. ***Recognize and account for bias.*** AI systems learn from human-created datasets, which may contain cultural, moral, political, or ideological bias. Use AI as a starting point, but not a final authority, especially in matters involving ethics, doctrine, or personal values.

7. ***Protect your moral agency.*** Use AI as a tool that supports your decisions, not as an authority that makes them. Be cautious of convenience replacing effort, or automation replacing thought. Guard against becoming passive or spiritually disengaged.

8. ***Use AI to enhance, not replace, creativity and effort.*** Let AI accelerate mundane tasks so you can invest more energy in creativity, inspired work, personal growth, and human connection. Avoid letting it do all the thinking,

writing, or creating for you, especially in important responsibilities.

9. ***Maintain healthy human relationships.*** Use technology to support, not supplant, connections with real people. Prioritize face-to-face conversations, family bonds, ministering, forgiveness, and shared experiences, which cannot be replicated by algorithms.

10. ***Set personal boundaries and use AI intentionally.*** Decide how, when, and why you will use AI. Establish limits on how much time you spend with it, what tasks you allow it to handle, and what emotional or intellectual space it occupies. Intentional use strengthens discipline and safeguards your agency.

## How to Select an AI Tool

AI tools are developing rapidly and constantly improving. The best tool today for a particular task may not be the best next week. Below are lists of some of the major tools on the market in early 2026. Many of them are free to use and others charge a small monthly fee. To discover the best tool for a specific task, just ask AI and it will guide you in the selection of a tool for your specific need.

### General Purpose Chatbots

Below is a list of the best general-purpose AI chatbots. Each is designed for broad, everyday uses, such as writing, research, conversation, coding, planning, and productivity.

- **ChatGPT (OpenAI).** Best overall general purpose chatbot. Highly accurate, mature GPT 5 and O1 models. Excellent reasoning, research abilities, and conversation quality. Strong for writing, coding, creativity, and general knowledge.

- **Gemini (Google).** Best for Google ecosystem and multimodal tasks. Deep integration with Gmail, Docs, Drive, and Calendar. Strong multimodal understanding (text, images, video). Great value because of integration and cloud storage.
- **Copilot (Microsoft).** Best for Windows and Office users. Seamless integration with Word, Excel, Outlook, and Teams. Uses GPT-5-based models. Strong productivity and workflow automation.
- **Perplexity.** Best for research, citations, and real time web answers. Strong search and chatbot hybrid. Excellent for fact checking, data lookups, and web insights.
- **Claude (Anthropic).** Best for writing, long form reasoning, and safety. Excellent at structured logic, long documents, and careful outputs. Very strong for writers, editors, and planning. Highly privacy focused.
- **Grok (X).** Best for real-time trending information inside X.com. Pulls directly from X for current events. More unfiltered and personality driven.
- **DeepSeek.** Best for reasoning and coding among open weight models. Very strong mathematical and logical reasoning. Useful for developers due to open weight availability.
- **Meta AI.** Best for social media ecosystem use. Integrated across Facebook, Instagram, and WhatsApp. Good for everyday general conversation and productivity.

- **Zapier Agents.** Best for automation and multi-step workflows. Connects chatbots to 8,000+ apps. Automates tasks, research, and workflows.
- **Poe (Quora).** Best multi model access in one place. Access ChatGPT, Claude, Gemini, Llama models, etc. Useful if you want many chatbots in a single interface.

**Image Generation (AI Art And Design)**

These tools create images from a text prompt.

- *Top AI image-generation tools:* Midjourney (known for stunning, artistic, highly stylized images; great for concept art, product visualizations, landscapes, and portraits), OpenAI DALL·E 3 (best for literal prompt accuracy and text understanding; creates clean, consistent, realistic images; excellent for illustrations, logos, and scenes with complex relationships), and Stable Diffusion (for art, product renders, photorealism, anime, and stylized work; fully open-source; endlessly customizable; massive ecosystem).
- *Other leading image-generation models:* Adobe Firefly (integrated across Photoshop and Illustrator; excellent for professional workflows, editing, inpainting, and text effects; safe for commercial-use), Google Imagen 2 (high-quality, photorealistic results; strong text accuracy and detail; integrated into Google products like Vertex AI), Ideogram (best-in-class text rendering inside images; great for posters, book covers, and thumbnail design), Leonardo.ai (feature-rich platform with workflows, styles, and training your own models; great

for product shots, fantasy art, and gaming assets), Runway Gen-2 (known for creative, cinematic, and experimental visuals; hybrid image + video generation platform), PixArt (open model with strong photorealism; popular in research and pro workflows), Playground AI (easy, free, browser-based; multiple models inside; very flexible editing and upscaling tools), BlueWillow (free, beginner-friendly alternative to Midjourney; makes stylized and fantasy artwork), Artbreeder (great for blending, morphing, genetic-style image creation; popular for characters and stylized portraits), Craiyon [formerly DALL·E Mini] (lightweight, simple generator; good for experimentation), NightCafe Creator (multi-model support, easy workflows; credits-based system for unlimited creative play), and Fotor AI Image Generator (simple, consumer-friendly generator; ideal for quick art, illustrations, and posters).

**Text-to-Video and Video Generation**

These tools turn scripts or prompts into finished videos.

- *Top professional-grade options:* Synthesia (best for training and explainer videos using AI avatars), LTX Studio (excellent for cinematic AI videos), OpenArt (great for artistic flexibility), and HeyGen (realistic AI avatars and corporate-style videos).
- *Simple, fast, template-driven tools:* Veed.io (easy web-based editing, plus AI text-to-video), InVideo (perfect

for quick social-media-ready videos), and Pictory (turns blogs and long content into short videos).

- *Feature-rich editors with AI assist:* PowerDirector (strong text-to-video, image-to-video, and AI avatars) and MyEdit (quick, browser-based text-to-video tool).

**Text Creation for Writing, Scripts, Ideas**

These tools are great for scripts, descriptions, captions, and brainstorming.

- ChatGPT (scriptwriting, captions, ideas, and keyword optimization)
- Canva Magic Write (social copy plus design integration)

**Content Creation**

Here are some of the best tools to help creators assemble and repurpose content quickly.

- AI Studios DeepBrain (2,000+ avatars, multilingual video creation)
- OpusClip (converts long videos into viral short clips automatically)
- Descript (edit video by editing text; great for podcasters and DIY projects)

# Home and Family

AI is changing home life the same way dishwashers and washing machines once did. Some people worry that AI may make home life impersonal or overly technical. But if used right, AI can handle routine tasks so you can focus on what matters most—your relationships.

The goal of using AI at home is to support your family and provide more time for real connection. Consider the list of ideas below on how to use AI at home.

1. **Meal Planning.** Input your family's favorite ingredients and dietary restrictions into an AI chatbot to generate a balanced weekly menu. Ask for a list that avoids allergens or sticks to a specific budget. Benefit: Reduces decision fatigue and supports family health.

2. **Recipe Suggestions.** Ask AI to suggest a meal based on five specific items in your fridge. Ask AI to suggest creative ways to "remix" a standard family recipe. Benefit: Minimizes food waste.

3. **Personalized Shopping.** Use AI to build grocery lists and organize it by section of the store. Ask for lists prioritizing seasonal or budget-friendly items. Benefit: Increases shopping efficiency.

4. **Smart Calendars.** Use AI-powered calendars to coordinate school, work, and church activities. Let the tool suggest better times when activities overlap. Benefit:

Helps the family communicate better, feel less stressed, and find more time for family connection.

5. **Reminders.** Set up automated alerts for recurring tasks using a smart assistant such as Alexa, Siri, or Google Assistant. Use voice commands for birthdays and school deadlines. Benefit: Keeps the household running smoothly.

6. **Chore Management.** Generate a fair chore rotation for children. Ask the AI to assign tasks based on age-appropriateness. Benefit: Simplifies household logistics.

7. **Finding Family Time.** Analyze your calendar to find timewasters. Ask AI to identify repetitive manual tasks that could be automated. Benefit: Frees up time for real connection.

8. **Homework Helper.** Use AI as a tutor for difficult school subjects. Ask the AI to "explain like I'm ten" to make complex concepts more accessible. Benefit: Supports children's learning without doing the work for them.

9. **Home Energy Audit.** Analyze smart thermostat data to find savings. Upload your utility usage data to identify patterns. Benefit: Reduces utility bills.

10. **Routine Automation.** Create "scripts" for your home environment. Set lights and temperature to adjust automatically to your routine. Benefit: Enhances home comfort and efficiency.

11. **Voice Assistance.** Manage tasks hands-free while multitasking. Dictate lists or ask questions while cooking

or cleaning, using Alexa, Siri, or Google Assistant. Ask the assistant to read you a book. Benefit: Allows you to remain productive while busy with manual labor.

12. **DIY Projects.** Research home repairs and other do-it-yourself projects. Give AI the model numbers or upload a photo to get specific step-by-step guidance for setting up devices or troubleshooting appliances that don't work. **Benefit:** Saves money on professional repairs.

13. **Family Activity Ideas.** Ask AI to generate ideas for family activities based on the family's preferences. Ask AI to develop a trivia game based on family history records. Benefit: Creates fun family activities.

# Work

Automation has replaced manual labor before, like when tractors replaced farm work in the early 1900s. Now, AI is changing office work in the same way. Instead of doing repetitive tasks, people can use AI to save time and focus on creative or more meaningful work.

Some people worry that AI will eliminate their jobs. Nvidia CEO Jensen Huang said, "You are not going to lose your job to an AI, but you are going to lose your job to someone who uses AI." (Message during the Milken Institute Global Conference, May 2025) AI is transforming the workplace. Embracing AI is no longer optional. It is essential for staying competitive.

In the 1980s, typists who resisted using word processing software were replaced by people who were willing to learn. Today, those who do not become proficient using AI tools may be replaced by those who do.

Below is a list of ideas on how to implement AI at work.

14. **Meeting Summaries.** Record and condense long discussions. Use AI to pull key points and decisions from a meeting recording. Benefit: Saves hours of manual notetaking and rewriting.

15. **Action Items.** Generate follow-up tasks instantly. Ask the AI to list "who does what" based on a transcript and email it to the team. Benefit: Ensures accountability.

16. **Scheduling.** Automate the "back-and-forth" of setting meetings. Use an AI agent to find mutually available times across different calendars. Benefit: Reduces administrative friction.

17. **Financial Processing.** Automate invoice and expense tagging. Set guidelines for an AI agent to flag unusual expenses for human review. Benefit: Can reduce weekly processing time from hours to seconds.

18. **Drafting Correspondence.** Create professional email templates. Provide context for the email and ask for a specific tone (e.g., firm but polite). Always review the text before sending it. Benefit: Saves time on routine communication.

19. **Automated Workflows.** Set trigger-based emails or events. Diagram workflows where an AI model can automatically act at specific trigger points. Benefit: Enables you to focus on more meaningful, creative work.

20. **Creating Excel Formulas.** Get instant help with data sheets. Describe what you want to calculate, and AI will provide the exact formulas. Benefit: Simplifies complex data analysis.

21. **Process Documentation.** Map out office workflows. Describe your daily routine to the AI and ask it to format it into a workflow diagram or even a training manual. Benefit: The workflow diagram could aid in automating workflows. The training manual can help new employees learn the job faster.

22. **Knowledge Sharing.** Synthesize what you've learned for the team. Use AI to summarize your successes and share them with coworkers to improve overall team efficiency. Benefit: Fosters a competitive and collaborative workplace.

23. **Creative Focus.** Offload repetitive tasks. Delegate data entry or basic scheduling to AI agents. Benefit: Allows you to spend more time on higher-level creative and problem-solving tasks.

24. **Strategic Planning.** Use industry patterns to stay competitive. Ask AI to find trends in market data to help you plan for the future. Benefit: Ensures you remain "the person who uses AI" and helps your company succeed.

# Religious Study

As you use AI technologies to enhance your gospel study, be careful that they don't become a substitute for inspiration from the Holy Ghost. Use technology to enhance but not replace the spiritual attentiveness and personal effort required to learn and progress in spiritual matters. Blessings come from wrestling with scripture, prayer, creativity, and personal revelation.

25. **Timeline Creation.** Visualize history. Ask AI to create a timeline of key events in Church history to better understand the sequence of events. Benefit: Gives you better perspective and context.

26. **Summarizing Articles.** Digest long religious texts quickly. Paste long articles or talks into a chatbot to extract the core message for study. Benefit: Makes learning efficient and focused.

27. **Topic Lists.** Identify themes within scriptures. Ask AI to list all instances of a specific theme within a certain book. Benefit: Enhances topical study.

28. **Question Generation.** Spark deeper reflection. Ask AI for deep questions based on a specific scripture to use for family discussion or personal prayer. Benefit: Turns information into "spiritual insights."

29. **Resource Recommendations.** Plan your next steps. Tell the AI your current study goals and ask for

recommended talks or scriptures to read next. Benefit: Personalizes your learning path.

30. **Comparing Sources.** Gain broader perspective. Use AI to compare verses across different Bible versions or summarize viewpoints from multiple reputable sources. Benefit: Helps discern truth and gain deeper insight.

31. **Cross-Referencing.** Link teachings across libraries. Quickly find connections between general conference talks and scripture verses. Benefit: Hastens the work of understanding the gospel.

# School

AI can enhance your educational experience when used wisely. Don't use it as a shortcut to replace genuine effort, but as a resource to strengthen your understanding.

32. **Step-by-Step Learning.** Break down hard subjects. Ask AI to divide a complex theory into manageable learning steps. Benefit: Prevents student overwhelm.

33. **Quiz Generation.** Test your knowledge. Provide your notes and ask the AI to create a practice quiz. Benefit: Reinforces understanding through active recall.

34. **Note Organization.** Clean up lecture notes. Paste your messy notes into AI and ask it to structure them into an outline or summary. Benefit: Improves study efficiency.

35. **Writing Feedback.** Use it as a virtual writing coach. Paste what you write into AI and ask for feedback on grammar, structure, and clarity. Benefit: Amplifies your own abilities.

36. **Practice Scenarios.** Roleplay for skill-building. Use AI to simulate public speaking or leadership situations to practice your responses. Benefit: Builds confidence in a safe environment.

37. **Factchecking.** Verify claims. Use AI as a starting point, then verify references against trustworthy, reputable sources. Benefit: Develops better discernment in a world of misleading information.

38. **Personalized Tutoring.** Learn at your own speed. Ask for explanations tailored specifically to your interests or "explain like I'm five." Benefit: Makes education personal and engaging.

39. **Literature Reviews.** Find patterns in research. Use AI to summarize themes across multiple academic papers for a thesis or project. Benefit: Speeds up the research phase.

40. **Time Management.** Create a custom study schedule. Input your exam dates and project deadlines to get a daily study plan. Benefit: Reduces procrastination.

# Personal Improvement and Learning

Outside of a formal school setting, make education a lifelong pursuit by constantly learning and improving yourself personally. AI can help you learn at your own pace, tailored to your interests and goals. Start by identifying what you want to learn and why, then consider the ideas below.

41. **Language Learning.** Practice conversation. Ask AI to speak with you in a new language to improve your fluency. Benefit: Prepares you to communicate with people who don't speak your language.

42. **Professional Skills.** Conduct a "gap analysis." Compare your resume with a dream job description and ask AI for a step-by-step learning plan to master the missing skills. Benefit: Keeps you competitive in the changing workforce.

43. **Progress Tracking.** Monitor your growth. Use AI to break goals into tasks and track your completion over time. Benefit: Increases motivation.

44. **Pattern Recognition.** Connect different fields. Ask AI how concepts from one hobby (e.g., gardening) apply to your career (e.g., project management). Benefit: Fosters innovation.

45. **Article Summarization.** Digest digital libraries. Use AI to summarize articles from online databases, then have it

ask you follow-up questions to delve deeper into topics that spark your interest. Benefit: Facilitates rapid learning.

46. **Critical Thinking.** Develop discernment. Verify AI answers with reputable sources to see where the AI might be "hallucinating." Benefit: Protects against misleading information.

47. **Reflective Journaling Prompts.** Deepen personal growth. Ask for daily prompts based on your current life goals to turn information into insight. Benefit: Enhances personal insight.

48. **Research Assistance.** Organize data points. Use AI to find and categorize key data for a project. Benefit: Streamlines complex research.

49. **Skill Practice.** Plan practice situations. Create a simulation to review principles you have recently learned in a leadership or technical course. Benefit: Reinforces useful knowledge.

50. **Evaluating Information Credibility.** Use AI to summarize arguments from a variety of reputable sources on a single topic to help you discern truth and avoid "hallucinations" or false online information. Ask AI to analyze the potential motivations of an organization providing online information to help identify bias. Benefit: Helps you find the truth.

51. **Small Business or Side-Hustle Brainstorming.** Use AI to brainstorm ideas for a home-based business or side hustle. It can also guide you through the process of

creating a business plan, marketing strategy, or customer service scripts. Benefit: Helps you provide for your family.

# Church Service and Community Volunteering

AI can help you fulfil your church callings and volunteer commitments by handling time-consuming administrative tasks. The goal is to save time on administrative tasks so you can focus more on personal service. Consider the following possibilities.

52. **Meeting Management.** Efficiently manage church or community meetings. AI can help you prepare agendas, summarize notes from the meeting, suggest follow-up discussion items, and send automated reminders for assignments. Benefit: Frees up time for personal connection and individual service.

53. **Community Needs Analysis.** Identify underserved areas. Analyze local news or community boards with AI to find needs that align with your talents. Benefit: Allows you to serve more effectively.

54. **Service Coordination.** Optimize volunteer time. Ask AI to suggest the best times for a group to meet and create plans to organize local resources. It can send reminders for planned visits or help with tracking service projects. Benefit: Improves community connectivity.

55. **Task Tracking.** Monitor project progress. Use AI to track the status of local service projects and ensure no tasks are forgotten. Benefit: Ensures project success.

56. **Event Promotion.** Create outreach materials. Use AI to design flyers and personalize social media posts to reach more people effectively. Benefit: Amplifies your voice in the community.

# Entertainment

As you use AI to find patterns and make tailored recommendations, you can discover new ways to enjoy your free time that match your unique goals. These tools help you handle the logistical aspects of entertainment, such as planning travel itineraries or organizing games, so you can focus on building relationships and enjoying real connection. When used thoughtfully, AI serves as a companion that amplifies your experiences and allows for more meaningful leisure time with those around you.

57. **Personalized Playlists.** Curate music. Ask AI for music recommendations based on a specific mood or activity. Benefit: Enhances daily atmosphere.

58. **Travel Planning.** Create itineraries. Ask for travel suggestions and points of interest tailored to your desires. Benefit: Simplifies trip organization.

59. **Game Design.** Engage family. Ask AI to create personalized puzzles or trivia games for family gatherings. Benefit: Provides wholesome recreation.

60. **Hobby Recommendations.** Discover new interests. List your skills and ask AI to suggest a new hobby you might enjoy. Benefit: Fosters personal growth.

61. **Interactive Fiction.** Play story games. Engage with AI-driven stories where your choices change the outcome of the stories. Benefit: Exercises creativity.

62. **Movie Suggestions.** Ask for movie recommendations based on genres you enjoy. Benefit: Tailors leisure time.

63. **Translation of Media.** Translate foreign works. Use Google Translate to translate books or films in real-time to enjoy global culture. Benefit: Broadens cultural understanding.

64. **Sports Analysis.** Track your teams. Analyze player stats for your local team or fantasy league to better understand and enjoy the game. Benefit: Adds depth to sports hobbies.

# Financial Management

Just as AI is transforming office work by automating repetitive tasks, it can also bring increased efficiency to your personal finances. By utilizing AI for routine processing, such as categorizing expenses and organizing receipts, you can reduce hours of manual labor to just a few seconds. These tools allow you to manage your resources with greater wisdom and purpose, ensuring you stay organized.

65. **Expense Management.** Track spending. Use AI to automatically tag and categorize spending and predict future expenses. Ask it to flag unusual account activity that deviates from your patterns. Benefit: Identifies areas to reduce costs.

66. **Subscription Management.** Stop "leaking" money. Ask AI to identify and help cancel unused recurring subscriptions. Benefit: Saves money.

67. **Tax Preparation.** Organize filings. Summarize income and deductible expenses for easier tax filing. Benefit: Reduces tax-season stress.

68. **Price Comparison.** Save on purchases. Use AI agents to find the best prices across multiple retailers. Benefit: Saves money.

69. **Bill Reminders.** Avoid late fees. Automate alerts for upcoming due dates so you never miss payments. Benefit: Saves money and protects your credit score.

70. **Savings Goals.** Plan major purchases. Create a step-by-step savings plan for a home or car using AI projections. Benefit: Helps reach financial milestones.

71. **Organize Receipts.** Catalog receipts. Use AI image recognition to digitally organize receipts for work or taxes. Benefit: Eliminates paper clutter.

# Health and Wellness

AI can enhance your well-being by tracking health patterns and providing insights into your physical condition. When used to manage routine tasks and interpret wellness data, these tools empower you to care for yourself and your family more effectively. This support allows you to focus on what matters most, ensuring you have the strength to efficiently do your part in your daily work and service.

72. **Dietary Tracking.** Analyze nutrition. Use AI to break down the nutritional content of your meals from a simple photo or description. Get personalized alerts to drink water based on your activity level. Benefit: Supports physical health.

73. **Fitness Planning.** Custom workouts. Create routines based on your specific fitness level, equipment, and goals. Benefit: Personalizes health growth.

74. **Sleep Analysis.** Improve rest. Use AI to interpret patterns in sleep data to find ways to get better rest. Benefit: Enhances overall quality of life.

75. **Medication Management.** Safety reminders. Set up smart reminders for doses and refills to ensure health consistency. Benefit: Prevents health errors.

76. **Health Research.** Understand conditions. Summarize complex medical articles to prepare for health decisions. Benefit: Empowers personal health stewardship.

77. **Doctor Appointments.** More productive doctor visits. Use AI to help formulate clear questions for your doctor based on your recent symptoms. Benefit: Makes medical consultations more effective.

78. **Stress Reduction.** Optimize for relaxation. Use AI to adjust your schedule to ensure time for mindfulness or gratitude journals. Benefit: Improves mental wellness.

# Family History and Heritage

AI is significantly speeding up the work of connecting families by reading old records, transcribing handwritten journals, and cataloging photos. These tools make family history easier to access, allowing you to learn from your ancestors and bring your heritage to life for the next generation.

79. **Transcribe and Translate Old Documents.** Unlock the past. Scan old handwritten letters and journals and ask AI to transcribe and translate them. Benefit: Speeds up family history work significantly.

80. **Photo Management.** Organize memories. Use AI to catalog, tag, sharpen, or colorize old family photographs. Benefit: Makes family history easier to access.

81. **Story Archiving.** Written narratives. Summarize family oral histories into readable written narratives. Benefit: Preserves the voices of ancestors.

82. **AI-Generated Family Timelines.** Create visual timelines. Ask AI to take birth dates, migration events, occupations, and major family milestones and automatically generate an illustrated timeline for an ancestor or for an entire family line. It could include historical context such as major world events happening at the same time. Benefit:

Helps family members visualize their heritage and understand how ancestors' lives fit into broader history.

83. **Ancestor Discovery.** Find connections. Use AI to identify missing connections in your family tree or find patterns in genealogical data. Benefit: Connects families across generations.

84. **Heritage Travel.** Plan meaningful trips. Research specific locations where your ancestors lived to plan a heritage trip. Benefit: Strengthens cultural identity.

85. **Immersive Family History Narrative.** Use AI to synthesize your family's genealogical data with deep historical context. Instead of merely viewing a birth record or a location, ask AI to create a story with cultural details, such as the clothing, local news, or typical daily routines of the town during the time your ancestor lived there. Benefit: Brings heritage to life for younger generations.

86. **Personal Histories.** Use AI to help document the life story of an older family member. Record interviews, then use AI to transcribe and organize them into a structured life history. Benefit: Provides an important historical document. Also, the process creates generational connections and insights.

# Creativity and Hobbies

AI is a remarkable tool meant to amplify human creativity and ingenuity, not to replace it. It can foster innovation by helping you brainstorm new ideas and explore artistic techniques. By offloading the routine aspects of your hobbies to AI, you can focus on the creative work that leads to personal growth and discovery.

87. **Creative Brainstorming.** Amplify your ingenuity. Use AI as a starting point to brainstorm and organize ideas for stories, books, or art. Benefit: Helps overcome "blank page" syndrome.

88. **Music Composition.** Songwriting aid. Generate soundtracks or melodies to help with amateur songwriting. Benefit: Fosters innovation.

89. **Gardening Advice.** Identify and care for plants. Use AI to identify plants from photos and get specific care instructions. Benefit: Improves home environment.

90. **Coding Assistance.** Learn technical skills. Ask AI to help write simple scripts or programs for personal hobbies. Benefit: Enhances technical literacy.

91. **Art Style Exploration.** Learn to paint. Ask AI to explain the techniques of various artists to improve your own skills. Benefit: Personal growth in the arts.

92. **Crafting Ideas.** Sustainable DIY. Suggest projects based specifically on leftover materials you have at home.

Benefit: Reduces waste and creates valuable do-it-yourself projects.

93. **Video Editing.** Organize memories. Use AI to automatically trim and organize clips from a family vacation into a movie. Benefit: Preserves family memories efficiently.

# Communication and Social Media

Innovations in modern communication help connect people across the globe. AI can help you create uplifting content and bridge language barriers, making it possible to communicate with more people than ever before. While these tools can amplify your voice and designs, they should be used to reflect your unique life experiences rather than substitute for them.

94. **Uplifting Content.** Share positive messages. Use AI to help create designs and text for social posts that share positive messages or personal stories. Personalize your online messages to reflect your unique life and voice. Benefit: Amplifies your voice in a creative, personal way.

95. **Message Clarity Check.** Avoid misinterpretation. Ask AI if an important email is polite and clear before sending it. Benefit: Enhances clear communication and protects relationships.

96. **Voice Assistance.** Empower individuals with speech or communication challenges by using AI tools that enable clearer expression and interaction. Benefit: Promotes inclusion and strengthens human connection.

97. **Language Translation.** Global communication. Use Google Translate to communicate with friends or

coworkers who speak other languages. Benefit: Expand your world of communication.

98. **Connection Suggestions.** Find your niche. Use AI to find groups or communities that share your specific interests. Benefit: Fosters community connection.

99. **Social Media Scheduling.** Plan for engagement. Plan the best times to post content to connect with friends and family. Benefit: Enhances social interaction.

100. **Drafting Speeches.** Structure outlines. Create outlines for wedding toasts or speeches in church or community meetings. Benefit: Ensures your message is organized, while allowing you to personalize with your own voice.

And the final idea:

101. **What else?** Ongoing innovation. Periodically ask AI what else it can help you with based on your previous interactions. Benefit: Ensures you are taking full advantage of technological discoveries.

www.ingramcontent.com/pod-product-compliance
Lightning Source LLC
LaVergne TN
LVHW010548100826
845148LV00013B/2655